ANIMAL TRACKS

of

KENTUCKY

&

TENNESSEE

Tamara Eder

with contributions from Ian Sheldon

LONE PINE

© 2002 by Lone Pine Publishing
First printed in 2002 10 9 8 7 6 5 4 3 2 1
Printed in Canada

THE PUBLISHER: LONE PINE PUBLISHING

1901 Raymond Avenue SW, Suite C	10145-81 Avenue
Renton, WA 98055	Edmonton, AB T6E 1W9
USA	Canada

Website: http://www.lonepinepublishing.com

National Library of Canada Cataloguing in Publication Data

Eder, Tamara, (date)
  Animal tracks of Kentucky and Tennessee

  Includes bibliographical references and index.
  ISBN 1-55105-319-5

  1. Animal tracks—Kentucky—Identification. 2. Animal tracks—Tennessee—
Identification. I. Sheldon, Ian, (date) II. Title.
QL768.E35 2001   591.47'9   C2001-910205-4

  Editorial Director: Nancy Foulds
  Editor: Volker Bodegom
  Proofreaders: Lee Craig, Genevieve Boyer
  Production Coordinator: Jen Fafard
  Design, Layout and Production: Volker Bodegom, Monica Triska
  Cover Design: Elliot Engley
  Cartography: Volker Bodegom
  Technical Contributor: Mark Elbroch
  Animal Illustrations: by Gary Ross, except for those by Ian Sheldon
      (p. 137), Kindrie Grove (p. 101), Ted Nordhagen (pp. 97, 127) and
      Ewa Pluciennik (p. 123).
  Track Illustrations: Ian Sheldon
  Cover Illustration: Eastern Chipmunk by Gary Ross
  Scanning: Elite Lithographers Ltd.

We acknowledge the financial support of the Government of Canada
through the Book Publishing Industry Development Program (BPIDP)
for our publishing activities.

PC: P4

CONTENTS

INTRODUCTION

If you have ever spent time with an experienced tracker, or perhaps a veteran hunter, then you know just how much there is to learn about the subject of tracking and just how exciting the challenge of tracking animals can be. Maybe you think that tracking is no fun, because all you get to see is the animal's prints. What about the animal itself—is that not much more exciting? Well, for most of us who don't spend a great deal of time in the beautiful wilderness of Kentucky and Tennessee, the chances of seeing the elusive Red Fox or the fun-loving River Otter are slim. The closest that we may ever get to some animals will be through their tracks, and they can inspire a very intimate experience. Remember, you are following in the footsteps of the unseen—animals that are in pursuit of prey, or perhaps being pursued as prey.

This book offers an introduction to the complex world of tracking animals. Sometimes tracking is easy. At other times it is an incredible challenge that leaves you wondering just what animal made those unusual tracks. Take this book into the field with you, and it can provide some help with the first steps to identification. Animals tracks and trails are this book's focus; you will learn to recognize subtle differences for both. There are, of course, many additional signs to consider, such as scat and food caches, all of which help you to understand the animal that you are tracking.

It takes many years to become an expert tracker. Tracking is one of those skills that grows with you as you acquire new knowledge in new situations. Most importantly, you will have an intimate experience with nature. You will learn the secrets of the seldom seen. The more you discover, the more you will want to know, and, by developing a good understanding of tracking, you will gain an excellent appreciation of the intricacies and delights of our marvelous natural world.

How to Use This Book

Most importantly, take this book into the field with you! Relying on your memory is not an adequate way to identify tracks. Track identification has to be done in the field or with detailed sketches and notes that you can take home. Much of the process of identification involves circumstantial evidence, so you will have much more success when standing beside the track.

This book is laid out in an easy-to-use format. Beginning on p. 126, a quick reference appendix to the tracks of all the animals illustrated in the book is a fast way to familiarize yourself with certain tracks. It guides you to the more informative descriptions of each animal.

Each description is illustrated with the appropriate footprints and the track patterns that it usually leaves. Although these illustrations are not exhaustive, they do show the tracks or groups of prints that you will most likely see. You will find a list of dimensions for the tracks, giving the general range, but there will always be extremes, just as there are with people who have unusually small or large feet. Under the category 'Size' (of animal), the 'greater-than' sign (>) is used when the size difference between the sexes is pronounced.

If you think that you may have identified a track, check the 'Similar Species' section. This section is designed to help you confirm your conclusions by pointing out other animals that leave similar tracks and by showing you ways to distinguish among them.

As you read this book, you will notice an abundance of words such as 'often,' 'mostly' and 'usually.' Unfortunately, tracking will never be an exact science; we cannot expect animals to conform to our expectations, so be prepared for the unpredictable.

Tips on Tracking

As you flip through this guide, you will notice clear, well-formed prints. Do not be deceived! It is a rare track that will ever show so clearly. For a good, clear print, the perfect conditions are slightly wet, shallow snow that is not melting, or slightly soft mud that is not actually wet. These conditions can be rare—most often you will be dealing with incomplete or faint prints where you cannot even really be sure of the number of toes.

Should you find yourself looking at a clear print, then the job of identification is much easier. There are a number of key features to look for: measure the length and width of the print, count the number of toes, check for claw marks and note how far away they are from the body of the print, and look for a heel mark. Keep in mind more subtle features, such as the spacing between the toes, whether or not they are parallel, and whether fur on the sole of the foot has made the print less clear.

When you are faced with the challenge of identifying an unclear print—or even if you think that you have made a successful identification from one print alone—look beyond the single footprint and search out others. Do not rely on the dimensions of one print alone, but collect measurements from several prints to get an average impression. Even the prints within one trail can show a lot of variation.

Try to determine which is the fore print and which is the hind, and remember that many animals are built very differently from humans, having larger forefeet than hind feet. Sometimes the prints will overlap, or they can be directly on top of one another in a direct register. For some animals, the fore and hind prints are pretty much the same.

Check out the pattern that the tracks make together in the trail and follow the trail for as many paces as is necessary for you to become familiar with the pattern. Patterns are very important and can be the distinguishing feature between different animals with otherwise similar tracks.

Follow the trail for some distance—it may give you some vital clues. For example, the trail may lead you to a tree, indicating that the animal is a climber, or it may lead down into a burrow. This part of tracking can be the most rewarding, because you are following the life of the animal as it hunts, runs, walks, jumps, feeds or tries to escape a predator.

Take into consideration the habitat. Sometimes habitat alone will allow you to distinguish very similar tracks—one species might be found on the riverbank, whereas another might be encountered just in dense forest.

Think about your geographical location, too, because some animals have a limited range. This consideration can rule out some species and help you with your identification.

Remember that every animal will at some point leave a print or trail that looks just like the print or trail of a completely different animal!

Finally, keep in mind that if you track quietly, you might catch up with the maker of the prints.

Eastern Spotted Skunk

Terms & Measurements

Some of the terms used in tracking can be rather confusing, and they often depend on personal interpretation. For example, what comes to your mind if you see the word 'hopping'? Perhaps you see a person hopping about on one leg, or perhaps you see a rabbit hopping through the countryside. Clearly, one person's perception of motion can be very different from another's. Some useful terms are explained below, to clarify what is meant in this book and, where appropriate, how the measurements given fit in with each term.

The following terms are sometimes used loosely and interchangeably—for example, a rabbit might be described as 'a hopper' and a squirrel as 'a bounder,' yet both leave the same pattern of prints in the same sequence.

Ambling: Fast, rolling walking.

Bounding: A gait of four-legged animals in which the two hind feet land simultaneously, usually registering in front of the fore prints. It is common in rodents and the rabbit family. 'Hopping' or 'jumping' can often be substituted.

hind prints

fore prints

Gait: Describes how an animal is moving at some point in time. Different gaits result in different observable trail characteristics.

Galloping: A gait used by animals with four legs of even length, such as dogs, moving at high speed, hind feet registering in front of forefeet.

Hopping: Similar to bounding. With four-legged animals it is usually indicated by tight clusters of prints, fore prints set between and behind the hind prints. A bird hopping on two feet creates a series of paired tracks along its trail.

Loping: Like galloping, but slower, with each foot falling independently and leaving a trail pattern that consists of groups of tracks in the sequence fore-hind-fore-hind, usually roughly in a line.

Mustelids (weasel family) often use ***2×2 loping***, in which the hind feet register directly on the fore prints. The resulting pattern has angled, paired tracks.

Running: Like galloping, but applied generally to animals moving at high speed. Also used for two-legged animals.

Trotting: Faster than walking, slower than running. The diagonally opposite limbs move simultaneously; that is, the right forefoot with the left hind, then the left forefoot with the right hind. This gait is the natural one for canids (dog family), short-tailed shrews and voles.

hind print

fore print

Canids may use ***side-trotting,*** a fast trotting in which the hind end of the animal shifts to one side. The resulting track pattern has paired tracks, with all the fore prints on one side and all the hind prints on the other.

Walking: A slow gait in which each foot moves independently of the others, resulting in an alternating track pattern. This gait is common for felines (cat family) and deer, as well as wide-bodied animals, such as bears and porcupines. The term is also used for two-legged animals.

Other Tracking Terms:

Dewclaws: Two small, toe-like structures set above and behind the main foot of most hoofed animals.

Direct Register: The hind foot falls directly on the fore print.

double register   direct register

Double Register: The hind foot registers and overlaps the fore print only slightly or falls beside it, so that both prints can be seen at least in part.

Dragline: A line left in snow or mud by a foot or the tail dragging over the surface.

dragline

Gallop Group: A track pattern of four prints made at a gallop, usually with the hind feet registering in front of the forefeet (see '*galloping*' for illustration).

Height: Taken at the animal's shoulder.

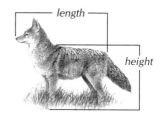

Length: The animal's body length from head to rump, not including the tail, unless otherwise indicated.

Metacarpal Pad: A small pad near the palm pad or between the palm pad and heel on the forefeet of bears and mustelids (members of the weasel family).

Print (also called '**track**'): Fore and hind prints are treated individually. Print dimensions given are 'length' (including claws—maximum values may represent occasional heel register for some animals) and 'width.' A group of prints made by each of the animal's feet makes up a track pattern.

Register: To leave a mark—said about a foot, claw or other part of an animal's body.

Retractable: Describes claws that can be pulled in to keep them sharp, as with felids (the cat family); these claws do not register in the prints. Foxes have semi-retractable claws.

Sitzmark: The mark left on the ground by an animal falling or jumping from a tree.

Straddle: The total width of the trail, all prints considered.

straddle

stride

Stride: For consistency among different animals, the stride is taken as the distance from the center of one print (or print group) to the center of the next one. Some books may use the term 'pace.'

Track: Same as '**print**.'

Track Pattern: The pattern left after each foot registers once; a set of prints, such as a gallop group.

Trail: A series of track patterns; think of it as the path of the animal.

Common Snipe

MAMMALS

River Otter

Elk

Fore and Hind Prints
Length: 3.2–5 in (8–13 cm)
Width: 2.5–4.5 in (6.5–11 cm)

Straddle
7–12 in (18–30 cm)

Stride
Walking: 16–34 in (40 85 cm)
Galloping: 3.3–8 ft (1–2.4 m)
Group length: to 6.3 ft (1.9 m)

Size (bull>cow)
Height: 4–5 ft (1.2–1.5 m)
Length: 6.5–10 ft (2–3 m)

Weight
500–1000 lb (230–450 kg)

gallop print | *walking*

18

ELK (Wapiti)
Cervus elaphus

Historically, the Elk ranged widely across North America. Although it has been absent from this region for well over 100 years, recent reintroduction efforts in both Kentucky and Tennessee mean that you may once again be finding this elegant animal's tracks here. Female Elk and young form social herds and consequently leave abundant tracks. Stags, however, prefer to go solo. Look for the Elk's tracks in the soft mud beside summer ponds.

The Elk leaves a neat alternating walking pattern of large, rounded prints, often in well-worn winter paths. The hind foot will sometimes double register slightly in front of the fore print. In deeper snow, or if an Elk gallops (with its toes spread wide), the dewclaws may register.

Similar Species: White-tailed Deer (p. 20) prints can be similar, but they are generally smaller. Feral Pig (p. 22) tracks show a shorter stride, and the prints have prominent dewclaws angled to the side.

White-tailed Deer

Fore and Hind Prints
Length: 2–3.5 in (5–9 cm)
Width: 1.6–2.5 in (4–6.5 cm)
Straddle
5–10 in (13–25 cm)
Stride
Walking: 10–20 in (25–50 cm)
Galloping: 6–15 ft (1.8–4.5 m)
Size (buck>doe)
Height: 3–3.5 ft (90–110 cm)
Length: to 6.3 ft (1.9 m)
Weight
120–350 lb (55–160 kg)

walking   *gallop group*

WHITE-TAILED DEER
Odocoileus virginianus

The keen hearing of this deer guarantees that it knows about you before you know about it. Frequently, all that we see is its conspicuous white tail in the distance as it gallops away, earning this deer the nickname 'Flagtail.' The adaptable White-tailed Deer can be found throughout the region in small groups at the edges of forests and in brushlands. It can be common around ranches and residential areas.

This deer's prints are heart-shaped and pointed. Its alternating walking track pattern shows the hind prints direct registered or double registered on the fore prints. When a deer gallops on a soft surface, the dewclaws register. This flighty deer gallops in the usual style, leaving hind prints ahead of fore prints, with toes spread wide for steadier, safer footing.

Similar Species: The Feral Pig (p. 22) makes similar tracks, but with a shorter stride, a wider gap between the toes and prominent dewclaws angled to the side.

Feral Pig

**Fore and Hind Prints
(with dewclaws)**
Length: 2.5–3 in (6.5–7.5 cm)
Width: 2.3 in (5.8 cm)
Straddle
5–6 in (13–15 cm)
Stride
Trotting: 16–20 in (40–50 cm)
Size (male>female)
Height: 3 ft (90 cm)
Length: 4.3–6 ft (1.3–1.8 m)
Weight
77–440 lb (35–200 kg)

trotting

FERAL PIG (Wild Pig, Wild Boar)
Sus scrofa

Descended from introduced European animals, the Feral Pig interbred with escaped Domestic Pigs (different in appearance but the same species). This sturdy beast of the dense undergrowth can be found in small, scattered populations in eastern Tennessee.

The Feral Pig's print shows two prominent, widely spaced toe marks, and usually, except on firm surfaces, a clear, pointed dewclaw mark off to each side. The hind print is slightly smaller than the fore print. Feral Pigs are keen foragers, so their tracks can often be numerous, especially when they travel in a group. The Feral Pig usually trots—a typical alternating track pattern shows a double register of hind print on fore print. Other signs are wallows and diggings.

Similar Species: White-tailed Deer (p. 20) prints can be similar, but they are slightly larger with a longer stride and a narrower gap between the toes; a deer's dewclaws may lightly register, but at the rear of the print, not the sides. Although the Domestic Pig is the same species, its tracks have a wider straddle and are less neat, often forming two separate lines.

Horse

Fore Print
(hind print is slightly smaller)
Length: 4.5–6 in (11–15 cm)
Width: 4.5–5.5 in (11–14 cm)

Straddle
2–7.5 in (5–19 cm)

Stride
Walking: 17–28 in (43–70 cm)

Size
Height: to 6 ft (1.8 m)

Weight
to 1500 lb (680 kg)

walking

HORSE
Equus caballus

Wilderness adventures on horseback are a popular activity, so you can expect horse tracks to show up almost anywhere.

Unlike any other animal in this book, the Horse has just one huge toe on each foot. This toe leaves an oval print with a distinctive 'frog' (V-shaped mark) at its base. If a Horse is shod, the horseshoe shows up clearly as a firm wall at the outside of the print. Not all horses are shod, so do not expect to see this outer wall on every horse print. A typical, unhurried horse trail shows an alternating walking pattern, with the hind prints registered on or behind the slightly larger fore prints. Horses are capable of a range of speeds—up to a full gallop—but most recreational horseback riders take a more leisurely outlook on life, preferring to walk their horses and soak up the beautiful views!

Similar Species: Mules (rarely shod) have smaller prints.

25

Black Bear

fore

hind

Fore Print
Length: 4–6.3 in (10–16 cm)
Width: 3.8–5.5 in (9.5–14 cm)

Hind Print
Length: 6–7 in (15–18 cm)
Width: 3.5–5.5 in (9–14 cm)

Straddle
9–15 in (23–38 cm)

Stride
Walking: 17–23 in (43–58 cm)

Size (male>female)
Height: 3–3.5 ft (90–110 cm)
Length: 5–6 ft (1.5–1.8 m)

Weight
200–600 lb (90–270 kg)

walking

BLACK BEAR
Ursus americanus

More common in Tennessee than Kentucky, the Black Bear has a much reduced and scattered range in remote forested areas. Do not expect to see its tracks in winter, when it sleeps deeply. Finding fresh bear tracks can be a thrill, but take care—the bear may be just ahead. Never underestimate the potential power of a surprised bear!

Black Bear prints somewhat resemble small human prints, but they are wider with claw marks. The small inner toe rarely registers. The forefoot's small heel pad often registers, and the hind print shows a big heel. The bear's slow walk results in a slightly pigeon-toed double register with the hind print on the fore print. More frequently, at a faster pace, the hind foot oversteps the forefoot. When a bear runs, the two hind feet register in front of the forefeet in an extended cluster. Along well-worn bear paths, look for 'digs' (patches of dug-up earth) and 'bear trees' whose scratched bark shows that this bear climbs.

Similar Species: No other species leaves similar tracks in this region.

Domestic Dog

fore

hind

Fore Print (hind print is smaller)
Length: 1–5.5 in (2.5–14 cm)
Width: 1–5 in (2.5–13 cm)

Straddle
1.5–8 in (3.8–20 cm)

Stride
Walking: 3–32 in (7.5–80 cm)
Loping to Galloping: to 9 ft (2.7 m)

Size
Very variable

Weight
Very variable

walking

loping to galloping

28

DOMESTIC DOG
Canis familiaris

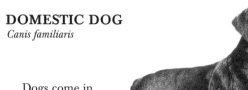

Dogs come in many shapes and sizes, from the tiny Chihuahua with its dainty feet to the robust and powerful Great Dane. Consequently, Domestic Dog tracks vary enormously. Dog ownership is high in many residential areas, and the popular pastime of dog walking can result in many dog tracks being left scattered about, especially if there is wet sand, mud or some snow.

The Domestic Dog's forefeet, which are much larger that the hind feet and support more of the animal's weight, leave the clearest tracks. When a dog walks, the hind prints usually register ahead of or beside the fore prints. As the dog moves faster, it trots and then lopes before it gallops. In a trot or lope pattern the prints alternate fore-hind-fore-hind, whereas a gallop group shows (from back to front) fore-fore-hind-hind.

Similar Species: Keep in mind that dog prints are usually found close to human tracks or activity. Fox (pp. 32–35) prints may be confused with small dog prints. Coyote (p. 30) prints—also similar but usually more oval and splaying less—will be in a more direct trail.

Coyote

fore

hind

Fore Print
(hind print is slightly smaller)
Length: 2.4–3.2 in (6–8 cm)
Width: 1.6–2.4 in (4–6 cm)

Straddle
4–7 in (10–18 cm)

Stride
Walking: 8–16 in (20–40 cm)
Trotting: 17–23 in (43–58 cm)
Galloping/Leaping:
      2.5–10 ft (0.8–3 m)

Size (female is slightly smaller)
Height: 23–26 in (58–65 cm)
Length: 32–40 in (80–100 cm)

Weight
20–50 lb (9–23 kg)

walking or trotting

gallop group

COYOTE
(Brush Wolf, Prairie Wolf)
Canis latrans

This widespread, adaptable canine prefers open grasslands or woodlands. On its own, with a mate or in a family pack, it hunts rodents and larger prey. If you find a coyote den—usually a wide-mouthed tunnel leading into a nesting chamber—do not bother the family or the female will have to move her pups to a safer location.

The hind print is slightly smaller than the oval fore print, and its less triangular heel pad rarely registers clearly. The claws of the two outer toes usually do not register. The Coyote typically walks or trots in an alternating pattern; the walk has a wider straddle, and the trotting trail is often very straight. When a Coyote gallops, its hind feet fall in front of its forefeet; the faster it goes, the straighter the gallop group. The Coyote's tail hangs down and may leave a dragline in loose material.

Similar Species: A Domestic Dog's (p. 28) less oval prints splay more, and its trail is erratic. Foot hairs blur Red Fox (p. 32) prints (usually smaller). Gray Fox (p. 34) prints are much smaller.

Red Fox

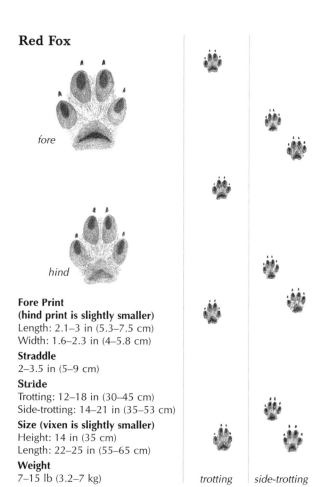

fore

hind

Fore Print
(hind print is slightly smaller)
Length: 2.1–3 in (5.3–7.5 cm)
Width: 1.6–2.3 in (4–5.8 cm)
Straddle
2–3.5 in (5–9 cm)
Stride
Trotting: 12–18 in (30–45 cm)
Side-trotting: 14–21 in (35–53 cm)
Size (vixen is slightly smaller)
Height: 14 in (35 cm)
Length: 22–25 in (55–65 cm)
Weight
7–15 lb (3.2–7 kg)

trotting   *side-trotting*

RED FOX
Vulpes vulpes

Very adaptable and intelligent, this beautiful and notoriously cunning fox is found throughout this region in a variety of habitats from forests to open areas.

Abundant foot hair allows just parts of the toes and heel pads to register, with no fine detail. The horizontal or slightly curved bar across the fore heel pad is diagnostic. A trotting Red Fox leaves a straight alternating trail— the hind print direct registers on the wider fore print. When the fox side-trots, its print pairs show the hind print falling to one side of the fore print in typical canid fashion. This fox gallops like the Coyote (p. 30). The faster the gallop, the straighter the gallop group.

Similar Species: Other canid prints lack the bar across the fore heel pad. Gray Fox (p. 34) prints are smaller. Domestic Dog (p. 28) prints can be of similar size, but they have a shorter stride and a less direct trail. Small Coyote prints are similar, but they have a wider straddle, and the toe marks are more bulbous.

Gray Fox

fore

hind

Fore Print
(hind print is slightly smaller)
Length: 1.3–2.1 in (3.3–5.3 cm)
Width: 1.1–1.5 in (2.8–3.8 cm)

Straddle
2–4 in (5–10 cm)

Stride
Walking/Trotting: 7–12 in (18–30 cm)

Size
Height: 14 in (35 cm)
Length: 21–30 in (53–75 cm)

Weight
7–15 lb (3.2–7 kg)

walking

GRAY FOX
Urocyon cinereoargenteus

   This small, shy fox is widespread, but it especially prefers woodlands and chaparral country. The Gray Fox is the only fox that climbs trees, which it does either for safety or to forage.

   The forefoot registers better than the smaller hind foot, and the hind foot's long, semi-retractable claws do not always register. The heel pads are often unclear—they sometimes show up as small, round dots. When it walks, this fox leaves a neat alternating track pattern. When it trots, its prints fall in pairs, with the fore print set diagonally behind the hind print. The Gray Fox's gallop group is like the Coyote's (p. 30).

Similar Species: The Red Fox (p. 32) has heel pads with a bar across them; its prints are generally larger and less clear (because of thick fur), its stride is longer, and its straddle is narrower. Coyote tracks are much larger. Domestic Cat (p. 38) and Bobcat (p. 36) prints lack claw marks and have larger, less symmetrical heel pads.

Bobcat

fore

hind

Fore Print
(hind print is slightly smaller)
Length: 1.8–2.5 in (4.5–6.5 cm)
Width: 1.8–2.5 in (4.5–6.5 cm)

Straddle
4–7 in (10–18 cm)

Stride
Walking: 8–16 in (20–40 cm)
Running: 4–8 ft (1.2–2.4 m)

Size (female is slightly smaller)
Height: 20–22 in (50–55 cm)
Length: 25–30 in (65–75 cm)

Weight
15–35 lb (7–16 kg)

walking

ambling to loping

BOBCAT
(Wildcat)
Lynx rufus

The widely distributed Bobcat, a stealthy and usually nocturnal hunter, is seldom seen. Very adaptable, it can leave tracks anywhere from wild mountainsides to chaparral and even in residential areas.

A walking Bobcat's hind feet usually register directly on its larger fore prints. As the Bobcat picks up speed, its trail becomes an ambling pattern of paired prints, the hind leading the fore. At even greater speeds, it leaves four-print groups in a lope pattern. The fore prints especially show asymmetry. The front part of the heel pad has two lobes and the rear part has three. The Bobcat's feet occasionally leave draglines.

Similar Species: A large Domestic Cat (p. 38) will make similar prints, but with a shorter stride and a narrower straddle. Domestic Dog (p. 28), Fox (pp. 32–35) and Coyote (p. 30) prints are narrower than they are long and show claw marks, and the fronts of their footpads are once-lobed; wild canid trails do not meander. Mink (p. 46) or Long-tailed Weasel (p. 48) prints that show four toes may also look similar.

Domestic Cat

fore

hind

Fore Print
(hind print is slightly smaller)
Length: 1–1.6 in (2.5–4 cm)
Width: 1–1.8 in (2.5–4.5 cm)

Straddle
2.4–4.5 in (6–11 cm)

Stride
Walking: 5–8 in (13–20 cm)
Loping/Galloping:
    14–32 in (35–80 cm)

Size (male>female)
Height: 20–22 in (50–55 cm)
Length with tail: 30 in (75 cm)

Weight
6.5–13 lb (3–6 kg)

walking

loping to galloping

DOMESTIC CAT
(House Cat)
Felis catus

The tracks of the familiar and abundant Domestic Cat can show up almost any place where there are people. Abandoned cats may roam farther afield; these 'feral cats' lead a pretty wild and independent existence. Domestic Cats come in many shapes, sizes and colors.

As with all felines, a Domestic Cat's fore print and slightly smaller hind print both show four toe pads. Its retractable claws, kept clean and sharp for catching prey, do not register. Cat prints usually show a slight asymmetry, with one toe leading the others. A Domestic Cat makes a neat alternating walking track pattern, usually in direct register, as one would expect from this animal's fastidious nature. When a cat picks up speed, it leaves clusters of four prints, the hind feet registering in front of the forefeet.

Similar Species: A small Bobcat (p. 36) may leave tracks similar to a very large Domestic Cat's. Domestic Dog (p. 28) and Fox (pp. 32–35) prints show claw marks.

Raccoon

fore

hind

Fore Print
Length: 2–3 in (5–7.5 cm)
Width: 1.8–2.5 in (4.5–6.5 cm)

Hind Print
Length: 2.4–3.8 in (6–9.5 cm)
Width: 2–2.5 in (5–6.5 cm)

Straddle
3.3–6 in (8.5–15 cm)

Stride
Walking: 8–18 in (20–45 cm)
Bounding: 15–25 in (38–65 cm)

Size (female is slightly smaller)
Length: 24–37 in (60–95 cm)

Weight
11–35 lb (5–16 kg)

walking

bounding group

RACCOON
Procyon lotor

The inquisitive Raccoon, which is found throughout this region, is adored by some people for its distinctive face mask, yet disliked for its boundless curiosity—often demonstrated with residential garbage cans. A good place to look for its tracks is near water at low elevations. The Raccoon likes to rest in trees. It usually dens up for the colder months.

The Raccoon's unusual print, showing five well-formed toes, looks like a human handprint; its small claws make dots. Its highly dexterous forefeet rarely leave heel prints, but its hind prints, which are generally much clearer, do show heels. The Raccoon's peculiar walking track pattern shows the left fore print next to the right hind print (or just in front) and vice versa. In mud or snow, a Raccoon may leave a direct-registering pattern instead. The Raccoon occasionally bounds, leaving clusters with the hind prints in front of the fore prints.

Similar Species: In wet mud, loose dirt or snow, unclear tracks by the Opossum (p. 42), River Otter (p. 44) or Woodchuck (p. 66) may look similar.

41

Opossum

fore

hind

Fore Print
Length: 2–2.3 in (5–5.8 cm)
Width: 2–2.3 in (5–5.8 cm)

Hind Print
Length: 2.5–3 in (6.5–7.5 cm)
Width: 2–3 in (5–7.5 cm)

Straddle
4–5 in (10–13 cm)

Stride
5–11 in (13–28 cm)

Size
Length: 2–2.5 ft (60–75 cm)

Weight
9–13 lb (4–6 kg)

walking

fast walking

OPOSSUM
Didelphis virginiana

This slow-moving, nocturnal marsupial is found throughout the region. It occupies many habitats and is quite tolerant of residential areas, but it prefers open woodland or brushland around waterbodies. Look for Opossum tracks in mud near the water or near roadkill (which Opossums like to eat, though many of them suffer the same fate as the carrion that they dine on).

The Opossum has two walking habits: the common alternating pattern, with the hind prints registering on the fore prints, and a Raccoon-like (p. 40) paired-print pattern, with each hind print next to the opposing fore print. The distinctive, long, inward-pointing thumb of the hind foot does not make a claw mark. The Opossum is an excellent climber, and its long, prehensile tail helps it maneuver amidst branches. However, contrary to popular myth, the Opossum does not regularly hang upside-down in trees by its tail.

Similar Species: Prints in which the distinctive thumbs do not show may be mistaken for a Raccoon's.

River Otter

fore

hind

Fore Print
Length: 2.5–3.5 in (6.5–9 cm)
Width: 2–3 in (5–7.5 cm)
Hind Print
Length: 3–4 in (7.5–10 cm)
Width: 2.3–3.3 in (5.8–8.5 cm)
Straddle
4–9 in (10–23 cm)
Stride
Loping: 12–27 in (30–70 cm)
Size
(female is two-thirds the size of male)
Length with tail: 3–4.3 ft (90–130 cm)
Weight
10–25 lb (4.5–11 kg)

loping (fast)

RIVER OTTER
Lontra canadensis

No animal knows how to have more fun than a River Otter. If you are lucky enough to watch one at play, you will not soon forget the experience. Widespread and well-adapted for the aquatic environment, this otter lives near water; an otter in the forest is usually on its way to another waterbody. Expect to see a wealth of evidence of an otter's presence along the riverbanks in its home territory. The River Otter loves to slide down riverbanks, leaving troughs nearly 1 foot (30 cm) wide.

In soft mud, the River Otter's five-toed feet, especially the hind ones, register evidence of webbing. The hind foot's inner toe is set slightly apart from the rest. If the forefoot's metacarpal pad registers, it lengthens the print. Very variable, otter trails usually show a typical mustelid 2×2 loping, but with faster gaits the tracks show groups of four and three prints. The thick, heavy tail often leaves a dragline.

Similar Species: Mink (p. 46) and Long-tailed Weasel (p. 48) prints are about half the size, and there will be no conspicuous tail drag.

Mink

fore

hind

Fore and Hind Prints
Length: 1.3–2 in (3.3–5 cm)
Width: 1.3–1.8 in (3.3–4.5 cm)
Straddle
2.1–3.5 in (5.3–9 cm)
Stride
Walking/Loping: 8–36 in (20–90 cm)
Size (female is slightly smaller)
Length with tail: 19–28 in (48–70 cm)
Weight
1.5–3.5 lb (0.7–1.6 kg)

2×2 loping

MINK

Mustela vison

The lustrous Mink, which is widespread throughout the region, prefers watery habitats surrounded by brush or forest. At home as much on land as in water, this nocturnal hunter can be exciting to track. Like the River Otter (p. 44), the Mink sometimes slides down riverbanks, carving out a trough up to 6 inches (15 cm) wide for an observant tracker to spot.

The Mink's fore print shows five (perhaps four) toes, with five loosely connected palm pads in an arc, but the hind print shows only four palm pads. The metacarpal pad of the forefoot rarely registers, but the furred heel of the hind foot may register, lengthening the hind print. The Mink prefers the typical mustelid 2×2 loping gait, which leaves consistently spaced, slightly angled double prints. Its diverse track patterns also include alternating walking, loping with three- and four-print groups (like the River Otter) and bounding (like a rabbit or hare, pp. 56–59).

Similar Species: The Long-tailed Weasel (p. 48) makes similar but generally smaller tracks. The River Otter's trail shows distinct tail draglines. Bobcat (p. 36) prints may at first glance resemble four-toed Mink prints.

Long-tailed Weasel

Fore and Hind Prints
Length: 1.1–1.8 in (2.8–4.5 cm)
Width: 0.8–1 in (2–2.5 cm)

Straddle
1.8–2.8 in (4.5–7 cm)

Stride
Bounding: 9.5–43 in (24–110 cm)

Size (male>female)
Length with tail: 12–22 in (30–55 cm)

Weight
3–12 oz (85–340 g)

2×2 loping

LONG-TAILED WEASEL
Mustela frenata

The Long-tailed Weasel, the largest of the two weasels in the region, is widely distributed. It is an active year-round hunter, with an avid appetite for rodents. Following this nimble creature's tracks can reveal much about its activities. A weasel trail may lead you up a tree or to the edge of water—weasels are accomplished swimmers. It might disappear into a rodent burrow if the weasel has pursued its occupant underground.

The usual weasel gait is a 2×2 lope, leaving a trail of paired prints. The Long-tailed Weasel's typical 2×2 lope shows an irregular stride—sometimes short and sometimes long—with no consistent behavior. Like the Mink (p. 46), this weasel may bound like a rabbit or hare (pp. 56–59).

Similar Species: The Least Weasel (p. 50), which occupies some of the same habitats, makes much smaller tracks. Mink tracks, though similar, are generally larger.

Least Weasel

Fore and Hind Prints
Length: 0.5–0.8 in (1.3–2 cm)
Width: 0.4–0.5 in (1–1.3 cm)

Straddle
0.8–1.5 in (2–3.8 cm)

Stride
Loping: 5–20 in (13–50 cm)

Size (male>female)
Length with tail: 6.5–9 in (17–23 cm)

Weight
1.3–2.3 oz (37–65 g)

2x2 loping

LEAST WEASEL
Mustela nivalis

Found primarily in eastern Kentucky and Tennessee, the Least Weasel is the smallest and rarest weasel in the region.

The Least Weasel's tracks may be found around wetlands and in open woodlands and fields. They are most evident in wet mud or moist sand, or if there is a dusting of snow. A weasel's light weight and small, hairy feet result in pad detail that is often unclear. Even with clear tracks, the inner (fifth) toe rarely registers.

Similar Species: A Long-tailed Weasel (p. 48) makes almost identical prints that are at least twice as large, and, given the animal's greater weight, they are likely to be more clear.

Striped Skunk

fore

hind

Fore Print
Length: 1.5–2.2 in (3.8–5.5 cm)
Width: 1–1.5 in (2.5–3.8 cm)

Hind Print
Length: 1.5–2.5 in (3.8–6.5 cm)
Width: 1–1.5 in (2.5–3.8 cm)

Straddle
2.8–4.5 in (7–11 cm)

Stride
Walking/Bounding:
 2.5–8 in (6.5–20 cm)

Size
Length with tail:
 20–32 in (50–80 cm)

Weight
6–14 lb (2.7–6.5 kg)

walking (fast)   *running*

STRIPED SKUNK
Mephitis mephitis

This striking skunk is notorious for its vile smell, and the lingering odor is often the best sign of its presence. Widespread throughout these two states, it thrives in diverse habitats at lower elevations. The Striped Skunk dens up in winter, coming out on warmer days and in spring.

Both the forefeet and hind feet have five toes. The long claws on the forefeet often register. The smooth palm pads and small heel pads leave surprisingly small prints. The Striped Skunk mostly walks—with such a potent smell for its defense, it rarely needs to run. Note that this skunk's trail rarely shows any consistent pattern, but an alternating walking pattern may be evident. The greater a skunk's speed, the more the hind foot oversteps the fore. If it runs, its trail consists of clumsy, closely set four-print groups. It occasionally drags its feet.

Similar Species: The Eastern Spotted Skunk (p. 54) makes smaller prints in a very random pattern. Mink (p. 46) or Long-tailed Weasel (p. 48) tracks will be farther apart than a skunk's. Skunk prints do not overlap.

Eastern Spotted Skunk

fore

hind

Fore Print
Length: 1–1.3 in (2.5–3.3 cm)
Width: 0.9–1.1 in (2.3–2.8 cm)

Hind Print
Length: 1.2–1.5 in (3–3.8 cm)
Width: 0.9–1.1 in (2.3–2.8 cm)

Straddle
2–3 in (5–7.5 cm)

Stride
Walking: 1.5–3 in (3.8–7.5 cm)
Bounding: 6–12 in (15–30 cm)

Size
Length: 13–25 in (33–65 cm)

Weight
0.6–2.2 lb (0.3–1 kg)

walking   *bounding*

EASTERN SPOTTED SKUNK
Spilogale putorius

This beautifully marked skunk, smaller than its striped cousin (p. 52), is declining in numbers throughout the region. It can live in diverse habitats—such as scrubland, forests and farmland—but it is a rare sight, because of its nocturnal habits, and because it dens up in winter, coming out only on warmer nights.

This skunk leaves a very haphazard trail as it forages for food on the ground. It occasionally climbs trees. Long claws on the forefeet often register, and the palm and heel may leave defined pad marks. Although this skunk rarely runs, when it does so it may bound along, leaving groups of four prints, hind in front of fore. It sprays only when truly provoked, so its powerful odor is less frequently detected than that of the Striped Skunk.

Similar Species: The Striped Skunk has larger prints and less scattered tracks with a shorter running stride (or it jumps); it does not climb trees.

Eastern Cottontail

fore

hind

Fore Print
Length: 1–1.5 in (2.5–3.8 cm)
Width: 0.8–1.3 in (2–3.3 cm)
Hind Print
Length: 3–3.5 in (7.5–9 cm)
Width: 1–1.5 in (2.5–3.8 cm)
Straddle
4–5 in (10–13 cm)
Stride
Hopping: 0.6–3 ft (18–90 cm)
Size
Length: 12–17 in (30–43 cm)
Weight
1.3–3 lb (0.6–1.4 kg)

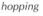

hopping

EASTERN COTTONTAIL
Sylvilagus floridanus

This abundant rabbit lives throughout Kentucky and Tennessee. Preferring brushy areas in grasslands and cultivated areas, it might be found in dense vegetation, hiding from predators such as the Coyote (p. 30) and the Bobcat (p. 36). Largely nocturnal, the Eastern Cottontail might be seen at dawn or dusk and on darker days.

As with other rabbits and hares, this cottontail's most common track pattern is a four-print triangular grouping, with the larger hind prints (which can appear pointed) in front of the fore prints (which may overlap). The hairiness of the toes will hide any pad detail. If you follow this rabbit's trail, you could be startled if it flies out from its 'form,' a depression in the ground in which it rests.

Similar Species: The Swamp Rabbit (*S. aquaticus*), found only in western parts of these states, makes similar but slightly larger prints. The Black-tailed Jackrabbit (p. 58) leaves much larger prints clusters and has a longer stride. Squirrel (pp. 70–75) tracks show a similar pattern, but with the fore prints more consistently side by side.

Black-tailed Jackrabbit

fore

hind

Fore Print
Length: 1.5–3 in (3.8–7.5 cm)
Width: 1.3–1.7 in (3.3–4.3 cm)

Hind Print
Length: 2.5–4 in (6.5–10 cm)
Length with heel: to 6 in (15 cm)
Width: 1.5–2.5 in (3.8–6.5 cm)

Straddle
4–7 in (10–18 cm)

Stride
Hopping: 5–10 ft (1.5–3 m)

Size
Length: 18–25 in (45–65 cm)

Weight
4–8 lb (1.8–3.6 kg)

hopping

BLACK-TAILED JACKRABBIT
Lepus californicus

Introduced to Kentucky, this athletic hare is quite rare, but it may be encountered in open and agricultural areas. The abundance of this hare fluctuates because of extensive hunting and continued introductions.

Both fore and hind prints show four toes. The hind foot may register a long heel when the hare walks slowly. When it hops, this hare creates print groups in a triangular pattern; as it speeds up, these print groups spread out considerably. Following a hare's trail could lead you to its 'form'—a depression where it rests—or to an urgent zigzag pattern that indicates where the hare fled from danger. With its strong hind legs, it is capable of leaping up to 20 feet (6 m) and running up to 35 mph (55 km/h) to avoid pursuers.

Similar Species: The Eastern Cottontail (p. 56) makes a much smaller print cluster and has a shorter stride. Coyote (p. 30) prints resemble heel-less jackrabbit prints, but the pattern is very different.

Nutria

fore

hind

Fore Print
Length: to 3 in (7.5 cm)
Width: to 3 in (7.5 cm)

Hind Print
Length: 4.5–6 in (11–15 cm)
Width: to 3.5 in (9 cm)

Straddle
to 7 in (18 cm)

Stride
Walking: to 8 in (20 cm)

Size (male>female)
Length with tail: 25–55 in (65–140 cm)

Weight
5–25 lb (2.3–11 kg)

walking

NUTRIA
(Coypu)
Myocastor coypus

Much larger than the Muskrat (p. 64), this rodent was introduced from South America by fur farmers. It has escaped into the wild, and there are now numerous colonies in the southern states, including Tennessee. With a voracious appetite for rice and sugar cane, the Nutria can be quite destructive to agricultural activities.

The webbing on the Nutria's strong hind feet is often evident in its prints. Look for claw marks, too. The large hind print shows five toes, with the innermost toe set farther back than the others. The much smaller fore print also shows five toes. The large, round, hairless tail often leaves a dragline. A Nutria's trail may lead you to a den in a riverbank, possibly a former Muskrat residence. Nearby, you might also notice large mats of vegetation— a Nutria's feeding platform.

Similar Species: Beaver (p. 62) prints are similar, but they show webbing between all the toes of the hind foot. Muskrat prints are smaller and show no webbing.

Beaver

fore

hind

Fore Print
Length: 2.5–4 in (6.5–10 cm)
Width: 2–3.5 in (5–9 cm)

Hind Print
Length: 5–7 in (13–18 cm)
Width: 3.3–5.3 in (8.5–13 cm)

Straddle
6–11 in (15–28 cm)

Stride
Walking: 3–6.5 in (7.5–17 cm)

Size
Length with tail: 3–4 ft (90–120 cm)

Weight
28–75 lb (13–34 kg)

walking

BEAVER
Castor canadensis

Few animals leave as many signs of their presence as the Beaver, North America's largest rodent and a common sight around water. Look for the conspicuous dams and lodges—capable of changing the local landscape—and the stumps of felled trees. Check trunks gnawed clean of bark for marks of the Beaver's huge incisors. Scent mounds marked with castoreum, a strong-smelling yellowish fluid that Beavers produce, also indicate recent activity.

Check the large hind prints for signs of webbing and broad toenails. The nail of the fourth toe usually does not register, and it is rare for all five toes on each foot to do so. Irregular foot placement in the alternating walking gait may produce a direct register or a double register. The Beaver's thick, scaly tail may mar its tracks, as can the branches that it drags about for construction and food. Repeated path use results in well-worn trails.

Similar Species: The Beaver's many signs, including large hind prints, minimize confusion. Nutria (p. 60) hind prints lack webbing between the fourth and fifth toes. Muskrat (p. 64) prints are much smaller.

Muskrat

fore

hind

Fore Print
Length: 1.1–1.5 in (2.8–3.8 cm)
Width: 1.1–1.5 in (2.8–3.8 cm)

Hind Print
Length: 1.6–3.2 in (4–8 cm)
Width: 1.5–2.1 in (3.8–5.3 cm)

Straddle
3–5 in (7.5–13 cm)

Stride
Walking: 3–5 in (7.5–13 cm)
Running: to 1 ft (30 cm)

Size
Length with tail: 16–25 in (40–65 cm)

Weight
2–4 lb (0.9–1.8 kg)

walking

MUSKRAT
Ondatra zibethicus

Like the Beaver (p. 62), this rodent is found throughout the region, wherever there is water. Beavers are very tolerant of Muskrats and even allow them to live in parts of their lodges. Active all year, the Muskrat leaves plenty of signs. It digs extensive networks of burrows, often undermining riverbanks, so do not be surprised if you suddenly fall into a hidden hole! Also look for small lodges in the water and beds of vegetation on which the Muskrat rests, suns and feeds in summer.

The small fifth (innermost) toe of the forefoot rarely registers. Stiff hairs, which aid in swimming, may create a 'shelf' around the five well-formed toes of the hind print. The common alternating walking pattern shows print pairs that alternate from side to side; the hind print is just behind the fore print or slightly overlaps it. In mud, sand or snow, a Muskrat's feet can drag, and its tail usually leaves a sweeping dragline.

Similar Species: Few animals share this water-loving rodent's habits. The Beaver and the Nutria (p. 60) both make larger tracks, with webbing registering on the hind prints.

Woodchuck

fore

hind

Fore and Hind Prints
Length: 1.8–2.8 in (4.5–7 cm)
Width: 1–2 in (2.5–5 cm)

Straddle
3.3–6 in (8.5–15 cm)

Stride
Walking: 2–6 in (5–15 cm)
Bounding: 6–14 in (15–35 cm)

Size (male>female)
Length with tail:
    20–25 in (50–65 cm)

Weight
5.5–12 lb (2.5–5.5 kg)

walking   *bounding*

WOODCHUCK
(Whistle Pig, Groundhog, Marmot)
Marmota monax

This robust member of the squirrel family is a common sight in open woodlands and adjacent open areas throughout much of the region. Always on the watch for predators, but not too troubled by humans, the Woodchuck never wanders far from its burrow. This marmot hibernates during winter and emerges in early spring; look for tracks in the mud around the burrow entrances in late spring.

A Woodchuck's fore print shows four toes, three palm pads and two heel pads (not always evident). The hind print shows five toes, four palm pads and two poorly registering heel pads. The Woodchuck usually leaves an alternating walking pattern, with the hind print registered on the fore print. When a Woodchuck runs from danger, it makes groups of four prints, hind ahead of fore.

Similar Species: A small Raccoon's (p. 40) bounding track pattern will be similar, but it will show five-toed fore prints.

Eastern Chipmunk

fore

hind

Fore Print
Length: 0.8–1 in (2–2.5 cm)
Width: 0.4–0.8 in (1–2 cm)

Hind Print
Length: 0.7–1.3 in (1.8–3.3 cm)
Width: 0.5–0.9 in (1.3–2.3 cm)

Straddle
2–3.2 in (5–8 cm)

Stride
Bounding: 7–15 in (18–38 cm)

Size
Length with tail: 7–10 in (18–25 cm)

Weight
2.5–5 oz (70–140 g)

bounding

EASTERN CHIPMUNK

Tamias striatus

Look for this delightful character throughout much of the region. The Eastern Chipmunk is found in a variety of habitats, from the dense forest floor to open areas near buildings. You are more likely to see or hear this rodent, which is highly active during summer, than to notice its tracks. This large chipmunk is happiest on the ground, but it will gladly climb sturdy oak trees to harvest juicy, ripe acorns. It hibernates in winter, waking up from time to time to eat.

Chipmunks are so light that their tracks rarely show fine details. The forefeet each have four toes and the hind feet have five. Chipmunks run on their toes, so the two heel pads of the forefeet seldom register; the hind feet have no heel pads. Their erratic track patterns, like those of many of their cousins, show the hind feet registered in front of the forefeet. A chipmunk trail often leads to extensive burrows.

Similar Species: No other chipmunks inhabit this region. Squirrels (pp. 70–75) usually make larger prints with a wider straddle, and they are more likely to make midwinter tracks. Mouse (pp. 84–87) tracks are smaller.

Eastern Gray Squirrel

fore

hind

Fore Print
Length: 1–1.8 in (2.5–4.5 cm)
Width: 1 in (2.5 cm)
Hind Print
Length: 2.3–3 in (5.8–7.5 cm)
Width: 1.1–1.5 in (2.8–3.8 cm)
Straddle
3.8–6 in (9.5–15 cm)
Stride
Bounding: 0.7–3 ft (22–90 cm)
Size
Length with tail: 17–20 in (43–50 cm)
Weight
14–25 oz (400–710 g)

bounding

EASTERN GRAY SQUIRREL

Sciurus carolinensis

This large, familiar squirrel can be a common sight in deciduous and mixed forests throughout this region, even in urban areas. Active all year, the Eastern Gray Squirrel can leave a wealth of evidence, especially in fall as it buries nuts for winter or if there is snow on the ground.

The Eastern Gray Squirrel leaves a typical squirrel track pattern when it runs or bounds. The hind prints fall slightly in front of the fore prints. A clear fore print shows four toes with sharp claws, four fused palm pads and two heel pads. The hind print shows five toes and four palm pads; if the full heel-length registers, it also shows two small heel pads.

Similar Species: Fox Squirrel (p. 72) prints are as big or larger. The Eastern Chipmunk (p. 68) and the Southern Flying Squirrel (p. 74) make smaller tracks in a similar pattern but with narrower straddles. Rabbits and hares (pp. 56–59) make longer track patterns, and their forefeet rarely register side by side when they run.

Fox Squirrel

fore

hind

Fore Print
Length: 1–1.9 in (2.5–4.5 cm)
Width: 1–1.7 in (2.5–4.3 cm)
Hind Print
Length: 2–3.3 in (5–7.5 cm)
Width: 1.5–1.9 in (3.8–4.5 cm)
Straddle
4–6 in (10–15 cm)
Stride
Bounding: 0.7–3 ft (22–90 cm)
Size
Length with tail: 18–28 in (45–70 cm)
Weight
1–2.4 lb (0.5–1.1 kg)

bounding

FOX SQUIRREL
Sciurus niger

This squirrel is much like the Eastern Gray Squirrel (p. 70), but it is larger and has a yellowish underside. It can be a common sight in deciduous forests with plenty of nut trees and in open areas or woodlands throughout the region. Its favorite feeding sites are indicated by piles of nutshells at tree bases. Active all year, the Fox Squirrel spends a lot of time foraging on the ground, often collecting nuts that it buried singly during the previous fall.

A clear fore print shows four toes with claws evident, four fused palm pads and two heel pads. The hind print shows five toes, four palm pads and sometimes a heel. When it runs or bounds, the Fox Squirrel makes a typical squirrel track, the hind prints slightly in front of the fore prints, and the prints in each pair roughly side by side.

Similar Species: The Eastern Gray Squirrel generally leaves smaller prints. Chipmunk (p. 68) and flying squirrel (p. 74) tracks are smaller and have narrower straddles. Rabbits and hares (pp. 56–59) make longer print patterns, and their fore prints rarely register side by side when they bound.

Southern Flying Squirrel

fore

hind

Fore Print
Length: 0.3–0.5 in (0.8–1.3 cm)
Width: 0.4 in (1 cm)
Hind Print
Length: 0.9–1.3 in (2.3–3.3 cm)
Width: 0.5 in (1.4 cm)
Straddle
2–2.5 in (5–6.5 cm)
Stride
Bounding: 7–22 in (18–55 cm)
Size
Length with tail: 8–10 in (20–25 cm)
Weight
1.5–3.2 oz (43–90 g)

bounding

SOUTHERN FLYING SQUIRREL
Glaucomys volans

This soft-furred brown acrobat is capable of long-distance gliding using the membranes that connect its forelegs and hind legs. Primarily nocturnal, the Southern Flying Squirrel is found in coniferous and mixed forests. In winter, up to 50 Southern Flying Squirrels may huddle together in a nest for warmth. Like other tree squirrels, this one does not truly hibernate.

Flying squirrels spend so much of their time in trees and gliding in the air that they make few tracks. Information from Mark Elbroch, a tracking expert, shows that when they travel on the ground the more characteristic pattern for Southern Flying Squirrels is a bound where the front tracks register in front of the rear tracks, and the straddle is narrower than the straddle of other squirrels.

Similar Species: Other squirrels (pp. 70–73) usually make larger prints and rarely leave sitzmarks, but with tracks in loose material it can be impossible to identify the squirrel species. Eastern Chipmunk (p. 68) tracks are smaller and have a narrower straddle.

Appalachian Woodrat

fore

hind

Fore Print
Length: 0.6–0.8 in (1.5–2 cm)
Width: 0.4–0.5 in (1–1.3 cm)
Hind Print
Length: 1–1.5 in (2.5–3.8 cm)
Width: 0.6–0.8 in (1.5–2 cm)
Straddle
2.3–2.8 in (5.8–7 cm)
Stride
Walking: 1.8–3 in (4.5–7.5 cm)
Bounding: 5–8 in (13–20 cm)
Size
Length with tail:
    14–17 in (35–43 cm)
Weight
13–16 oz (370–450 g)

walking   *bounding*

APPALACHIAN WOODRAT
(Allegheny Woodrat)
Neotoma magister

This nocturnal woodrat thrives in the rocky areas in the central and eastern parts of the region, but it feeds on foliage, seeds, ferns and fungi and therefore requires nearby vegetation. The trail of the Appalachian Woodrat might lead you to a distinctive mass of a nest, most often in a crevice, shrub or burrow.

Four toes show on the fore print and five on the hind. The short claws rarely register. A woodrat often walks in an alternating fashion, with the hind print direct registering on the fore print. This woodrat frequently bounds as well, leaving a pattern of four prints, with the larger hind prints in front of the diagonally placed fore prints. The stride tends to be short relative to the size of the prints.

Similar Species: The nearly identical Eastern Woodrat (*N. floridana*) lives in the west. The Norway Rat (p. 78) has similar prints, but it is usually found close to human activity. Hispid Cotton Rat (p. 80) tracks are usually smaller and have a narrower straddle. Indistinct squirrel (pp. 70–75) prints are similar but usually larger. Woodchuck (p. 66) prints are similar but much larger.

Norway Rat

fore

hind

Fore Print
Length: 0.7–0.8 in (1.8–2 cm)
Width: 0.5–0.7 in (1.3–1.8 cm)

Hind Print
Length: 1–1.3 in (2.5–3.3 cm)
Width: 0.8–1 in (2–2.5 cm)

Straddle
2–3 in (5–7.5 cm)

Stride
Walking: 1.5–3.5 in (3.8–9 cm)
Bounding: 9–20 in (23–50 cm)

Size
Length with tail: 13–19 in (33–48 cm)

Weight
7–18 oz (200–510 g)

walking

NORWAY RAT
(Brown Rat)
Rattus norvegicus

Active both day and
night, this despised rat is widespread almost anywhere that humans
have decided to build their homes. Not entirely
dependent on people, it may live in the wild as well.

The fore print shows four toes, and the hind print
shows five. When it bounds, this colonial rat leaves four-
print groups, with the hind prints in front of the diagonally placed fore prints. Sometimes one of the hind feet
direct registers on a fore print, creating a three-print
group. This rat more commonly leaves an alternating
walking pattern with the larger hind prints close to or
overlapping the fore prints; the hind heel does not show.
The tail often leaves a dragline in loose material. Rats
live in groups, so you may find many trails together,
often leading to their 5-inch (2-cm) wide burrows.

Similar Species: The Black Rat (*R. rattus*) and the Hispid Cotton Rat (p. 80) both leave similar tracks. Woodrat
(p. 76) tracks may be similar, but woodrats rarely associ-
ate with human activity, though they do live in abandoned buildings. Mouse (pp. 84–87) prints are much
smaller. The Eastern Chipmunk (p. 68) usually bounds,
and its tracks are often smaller.

Hispid Cotton Rat

fore

hind

Fore Print
Length: 0.5–0.7 in (1.3–1.8 cm)
Width: 0.5–0.7 in (1.3–1.8 cm)

Hind Print
Length: 0.6–1 in (1.5–2.5 cm)
Width: 0.6–0.8 in (1.5–2 cm)

Straddle
1.3–1.5 in (3.3–3.8 cm)

Stride
Walking: 1.3 in (3.3 cm)

Size
Length with tail: 8–14 in (20–35 cm)

Weight
2.8–7 oz (80–200 g)

walking

HISPID COTTON RAT
Sigmodon hispidus

The Hispid Cotton Rat, which is found in Tennessee and southern Kentucky, makes itself unpopular by eating valuable crops. Though keen on devouring almost anything green, it prefers grassy fields. It stays close to home, and consequently the rat's little runways clearly mark its routes to favored feeding sites.

The fore prints show four toes, but the larger hind prints usually show five. The heel of the hind foot will not always register, especially if the rat is moving fast. This medium-sized rodent leaves a typical walking track pattern in which the hind print registers on and slightly behind the fore print. Watch for its nests, which are balls of woven grass, and the small piles of cut grass it collects.

Similar Species: The Norway Rat (p. 78) and the Black Rat (*Rattus rattus*) also make similar tracks, but these two rats are usually found close to human activity. Woodrat (p. 76) tracks are usually larger and have a wider straddle.

Woodland Vole

fore

hind

Fore Print
Length: 0.5 in (1.3 cm)
Width: 0.5 in (1.3 cm)

Hind Print
Length: 0.6 in (1.5 cm)
Width: 0.5–0.8 in (1.3–2 cm)

Straddle
1.3–2 in (3.3–5 cm)

Stride
Walking/Trotting: 0.8 in (2 cm)
Bounding: 2–6 in (5–15 cm)

Size
Length with tail:
    4–5.5 in (10–14 cm)

Weight
0.8–1.3 oz (23–37 g)

walking

*bounding
(in snow)*

WOODLAND VOLE
(Pine Vole)
Microtus pinetorum

Distinguishing a vole track from the tracks of other small mammals in the region can be challenging. If you do spot a vole track, the most likely candidate is the Woodland Vole. It lives in a variety of habitats throughout the two states.

When clear (which is seldom), vole fore prints show four toes, and hind prints show five. A vole's walk and trot both leave a paired alternating track pattern with a hind print occasionally direct registered on a fore print. Voles usually opt for a faster bounding; the resulting print pairs show the hind prints registered on the fore prints. This vole lopes quickly across open areas, creating a three-print track pattern. In summer, well-used vole paths appear as little runways in the grass. Look for the distinctive piles of cut grass on runways near their ground nests. The bark at the bases of shrubs may show tiny teeth marks left by gnawing.

Similar Species: Three other less common voles whose similar tracks you may find are the Prairie Vole (*M. ochrogaster*), the Meadow Vole (*M. pennsylvanicus*) and the Rock Vole (*M. chrotorrhinus*). Mouse (pp. 84–87) bounding tracks show four-print groups.

White-footed Mouse

hind

fore

bounding group

Fore Print
Length: 0.3–0.4 in (0.8–1 cm)
Width: 0.3–0.4 in (0.8–1 cm)

Hind Print
Length: 0.3–0.6 in (0.8–1.5 cm)
Width: 0.3–0.4 in (0.8–1 cm)

Straddle
1.4–1.8 in (3.5–4.5 cm)

Stride
Bounding: to 12 in (30 cm)

Size
Length with tail:
5–8 in (13–21 cm)

Weight
0.4–1.5 oz (10–43 g)

bounding

bounding (in snow)

WHITE-FOOTED MOUSE

Peromyscus leucopus

The highly adaptable White-footed Mouse—one of the region's most abundant mammals—lives anywhere from cultivated land to wooded areas. It is seldom seen, because it is nocturnal. During inclement weather this mouse may enter buildings, where it will stay active.

In perfect soft mud, the fore prints each show four toes, three palm pads and two heel pads, and the hind prints show five toes and three palm pads; the hind heel pads rarely register. Bounding tracks, most noticeable in soft mud, sand or snow, show the hind prints in front of the fore prints. In sand the prints may merge to look like larger pairs of prints; tail drag will be evident. This mouse's trail may lead up a tree or down into a burrow.

Similar Species: Many less common species of mice have near-identical tracks. The House Mouse (*Mus musculus*), with very similar tracks, associates more with humans. Meadow Jumping Mouse (p. 86) prints show long, thin toes. Voles (p. 82) tend to trot and have a much shorter bounding track pattern. The Eastern Chipmunk (p. 68) has a wider straddle. Shrews (p. 88) have a narrower straddle.

Meadow Jumping Mouse

fore

hind

Fore Print
Length: 0.3–0.5 in (0.8–1.3 cm)
Width: 0.3–0.5 in (0.8–1.3 cm)
Hind Print
Length: 0.5–1.3 in (1.3–3.3 cm)
Width: 0.5–0.7 in (1.3–1.8 cm)
Straddle
1.8–1.9 in (4.5–4.8 cm)
Stride
Bounding: 7–18 in (18–45 cm)
In alarm: 3–6 ft (90–180 cm)
Size
Length with tail: 7–9 in (18–23 cm)
Weight
0.6–1.3 oz (17–35 g)

bounding

MEADOW JUMPING MOUSE
Zapus hudsonius

Congratulations if
you find and successfully
identify the tracks of the
Meadow Jumping Mouse!
Although it is abundant throughout most of the region,
its preference for grassy meadows and dense undergrowth—and its long, deep winter hibernation (about
six months!)—makes locating tracks very difficult.

Jumping mouse tracks are distinctive if you do find
them. The two smaller forefeet register between the long
hind prints; the long heels do not always register, and
some prints show just the three long middle toes. The
toes on the forefeet may splay so much that the side toes
point backward. When they bound, jumping mice make
short leaps. The tail may leave a dragline in soft mud or
snow. A more abundant sign of this rodent is the clusters
of grass stems, cut to about 5 inches (13 cm) long, which
are found lying in meadows.

Similar Species: White-footed Mouse (p. 84) tracks
may have the same straddle. Heel-less hind prints may
be mistaken for a vole's (p. 82) or a small bird's (p. 112)
or even an amphibian's (pp. 114–19).

Southeastern Shrew

hind

fore

bounding group

Fore Print
Length: 0.2 in (0.5 cm)
Width: 0.2 in (0.5 cm)

Hind Print
Length: 0.6 in (1.5 cm)
Width: 0.3 in (0.8 cm)

Straddle
0.8–1.3 in (2–3.3 cm)

Stride
Bounding: 1.2–3 in (3–7.5 cm)

Size
Length with tail: 2.3–4.5 in (7–11 cm)

Weight
0.07–0.14 oz (2–4 g)

bounding

SOUTHEASTERN SHREW

Sorex longirostris

Although several species of tiny, frenetic shrews are found in Kentucky and Tennessee, the widespread and adaptable Southeastern Shrew is a likely candidate if you find tracks. This shrew prefers moist fields, marshes, bogs or woodlands, but it can also be found in higher and drier fields.

In its energetic and unending quest for food, a shrew usually leaves a four-print bounding pattern, but it may slow to an alternating walking gait. The individual prints in a group are often indistinct, but in mud or shallow, wet snow you can even count the five toes on each print. In loose sand, dirt or light snow, a shrew's tail often leaves a dragline. A shrew's trail may disappear down a burrow.

Similar Species: The Smoky Shrew (*S. fumeus*) and the Least Shrew (*Cryptotis parva*) are two of the species that make similar prints. A mouse's (pp. 84–87) fore prints will show four toes.

Eastern Mole

a molehill of the Eastern Mole

some molehills and ridges of the Eastern Mole

Size
Length with tail: 3.5–8.5 in (9–22 cm)
Weight
3–5 oz (85–140 g)

EASTERN MOLE

Scalopus aquaticus

This soft-furred resident of the underworld is one of only two moles that you might encounter in this region. It can leave a wealth of evidence that indicates its presence, usually in pastures or open woodlands, especially where the soil is light and moist and easy to burrow in.

Moles, which seldom emerge from their subterranean environment, create an extensive network of burrows through which they forage. These burrows are sometimes marked by ridges on the surface, though most of us are more familiar with the hills that form where the mole gets rid of excess soil from its burrows and for which moles are frequently considered to be pests (they mess up fine lawns). When rains moisten the soil and bring worms to the surface, it can be entertaining to watch the earth twitch and rise up as the mole satisfies its voracious appetite.

Similar Species: The Hairy-tailed Mole (*Parascalops breweri*), which lives in the eastern parts of these states, makes similar hills.

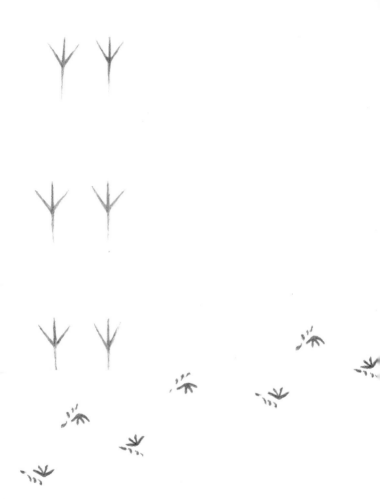

BIRDS, AMPHIBIANS & REPTILES

A guide to the animal tracks of Kentucky and Tennessee is not complete without some consideration of the birds, amphibians and reptiles found in these states.

Several bird species have been chosen to represent the main types common to this region, but remember that individual bird species are not easily identified by track alone. Bird tracks are often abundant—the shores of lakes and streams are very reliable places to find them—especially in wet sand, mud or snow, which can hold a clear print for a long time. The sheer number of tracks made by shorebirds and waterfowl can be astonishing. Although some bird species prefer to perch in trees or soar across the sky, it can be entertaining to follow the tracks of birds that spend a lot of time on the ground. They can spin around in circles and lead you in all directions. The trail may suddenly end as the bird takes flight, or it might terminate in a pile of feathers, the bird having fallen victim to a predator.

Many amphibians and turtles depend on moist environments, so look in the soft mud along the shores of lakes and ponds for their distinctive tracks. You may be able to distinguish frog tracks from toad tracks, because these two amphibians generally move differently, but it can be very difficult to identify the species. Reptiles thrive and outnumber the amphibians in drier environments, but they seldom leave good tracks, except in mud or perhaps in sand or snow. Snakes leave distinctive body prints.

Mallard

Print
Length: 2–2.5 in (5–6.5 cm)
Straddle
4 in (10 cm)
Stride
to 4 in (10 cm)
Size
23 in (58 cm)

MALLARD
Anas platyrhynchos

male

female

   This dabbling duck is common in open areas near lakes and ponds. The male is an especially familiar sight with its striking green head. The Mallard's webbed feet leave prints that can often be seen in abundance along the muddy shores of just about any waterbody, including those in urban parks.

   The webbed foot of the Mallard has three long toes that all point forward. Although the toes register well, the webbing between the toes does not always show in the print. The inward-pointing feet give the Mallard a pigeon-toed appearance and perhaps account for its waddling gait, a characteristic for which ducks are known.

Similar Species: Many waterfowl, such as other ducks, as well as the Herring Gull (p. 96), leave similar prints. Exceptionally large prints were likely made by geese (various species) or by a swan (*Cygnus* spp.).

Herring Gull

Print
Length: 3.5 in (9 cm)
Straddle
4–6 in (10–15 cm)
Stride
4.5 in (11 cm)
Size
Length: 23–25 in (58–65 cm)

HERRING GULL
Larus argentatus

The Herring Gull, with its long wings and webbed toes, is a strong long-distance flier as well as an excellent swimmer. Increasingly common in the region, this bird is concentrated in great numbers near waterbodies and garbage dumps.

Gulls leave slightly asymmetrical tracks that show three toes. They have claws that register outside the webbing, and the claw marks are usually attached to the footprint. Most gulls have quite a swagger to their gait, and they leave a trail with the tracks turned strongly inward.

Similar Species: Gull species cannot be reliably identified by track alone, but smaller species have conspicuously smaller tracks. Mallard (p. 94) and other duck tracks are often difficult to distinguish from gull tracks. Geese (various species) make larger prints.

Great Blue Heron

Print
Length: to 6.5 in (17 cm)

Straddle
8 in (20 cm)

Stride
9 in (23 cm)

Size
4.2–4.5 ft (1.3–1.4 m)

GREAT BLUE HERON
Ardea herodias

The regal and graceful image of this large heron symbolizes the precious wetlands in which it patiently hunts for food. Usually still and statuesque as it waits for a meal to swim by, the Great Blue Heron will occasionally walk from time to time, perhaps to find a better hunting location. Look for its large, slender tracks along the banks or mudflats of waterbodies.

Not surprisingly, a bird that lives and hunts with such precision walks in a similar fashion, leaving straight tracks that fall in a nearly straight line. Look for the slender rear toe in the print.

Similar Species: Other herons and egrets (various species) have similar prints that vary in size relative to the size of the bird.

Common Snipe

Print
Length: 1.5 in (3.8 cm)
Straddle
to 1.8 in (4.5 cm)
Stride
to 1.3 in (3.3 cm)
Size
11–12 in (28–30 cm)

COMMON SNIPE
Gallinago gallinago

This short-legged character is a resident of marshes and bogs, where its neat prints can often be seen in mud. Snipes are quite secretive when on the ground, and so you may be surprised if one suddenly flushes out from beneath your feet. If there is a Common Snipe in the air, you may hear an eerie whistle if it dives from the sky.

The Common Snipe's neat prints show four toes, including a small rear toe that points inward. The bird's short legs and stocky body give it a very short stride.

Similar Species: Many shorebirds, including the Spotted Sandpiper (p. 102), leave similar tracks.

Spotted Sandpiper

Print
Length: 0.8–1.3 in (2–3.3 cm)

Straddle
to 1.5 in (3.8 cm)

Stride
Erratic

Size
7–8 in (18–20 cm)

SPOTTED SANDPIPER

Actitis macularia

The bobbing tail of the Spotted Sandpiper is a possible sight on the shores of lakes, rivers and streams. You will usually find just one of these territorial birds in any given location. These sandpipers may be declining because of habitat degradation. With its excellent camouflage, likely the first sight that you will see of this bird will be when it flies away, its fluttering wings close to the surface of the water.

As a sandpiper teeters up and down on the shore, it leaves trails of three-toed prints that show a very small fourth toe that faces off to one side at an angle. Sandpiper tracks can have an erratic stride.

Similar Species: All sandpipers and plovers, including the common Killdeer (*Charadrius vociferus*), leave similar tracks, although there is much diversity in size. The Common Snipe (p. 100) makes similar but larger tracks.

Ruffed Grouse

Print
Length: 2–3 in (5–7.5 cm)
Straddle
0.5–2 in (1.3–5 cm)
Stride
Walking: 3–6 in (7.5–15 cm)
Size
15–19 in (38–48 cm)

RUFFED GROUSE
Bonasa umbellus

This ground-dweller prefers the quiet seclusion of coniferous forests, where its excellent camouflage usually affords it good protection. Although you might find its tracks in mud, light snow much improves your chances of finding them. If you follow a Ruffed Grouse trail quietly, you may be startled when the bird bursts from cover almost beneath your feet.

The three thick front toes leave very clear impressions, but the short rear toe, which is angled off to one side, does not always show up so well. This bird's neat, straight trail appears to reflect its cautious approach to life on the forest floor.

Similar Species: The Northern Bobwhite (*Colinus virginianus*) leaves similar but smaller tracks. The Wild Turkey (*Meleagris gallopavo*) leaves similar but much larger tracks.

Great Horned Owl

Strike
Width: to 3 ft (90 cm)
Size
22 in (55 cm)

GREAT HORNED OWL
Bubo virginianus

Often seen resting
quietly in trees by day,
this wide-ranging owl
prefers to hunt at night.
The mark that this accom-
plished hunter may leave in
dust, sand or snow can be quite
a sight if it registers well. You
might stumble across such a
'strike' and guess that the owl's
target could have been a vole
(p. 82) or mouse (pp. 84–87)
scurrying around on the ground. If
you are a really lucky tracker, you might be following
the trail of a small animal to find that it abruptly ends
with this strike mark, where the animal has been seized.

The owl strikes with its talons, leaving scratch marks
that are occasionally surrounded by the imprints of wings
and tail feathers. These imprints result as the owl strug-
gles to take off with possibly heavy prey. Not the most
graceful of walkers, it prefers to fly away from the scene.

Similar Species: Other large birds, such as hawks, can
also leave strike marks, but the marks usually have less
rounded, more distinct feather imprints.

American Crow

Print
Length: 2.5–3 in (6.5–7.5 cm)
Straddle
1.5–3 in (3.8–7.5 cm)
Stride
Walking: 4 in (10 cm)
Size
16 in (40 cm)

AMERICAN CROW
Corvus brachyrhyncos

The black silhouette of the American Crow can be a common sight in a variety of habitats. A crow will frequently come down to the ground and contentedly strut around. Its loud *caw* can be heard from quite a distance. Crows can be especially noisy when they are mobbing an owl or a hawk.

The American Crow typically leaves an alternating walking track pattern. Its prints show three sturdy toes pointing forward and one toe pointing backward. When a crow is in need of greater speed, perhaps for take-off, it bounds along, leaving irregular pairs of diagonally placed prints with a longer stride between each pair.

Similar Species: Other corvids, such as jays (various species), also spend a lot of time on the ground and make similar tracks that vary in size according to the size of the bird.

Northern Flicker

Print
Length: 1.8 in (4.5 cm)

Straddle
1–1.5 in (2.5–3.8 cm)

Stride
Hopping: 1.5–5 in (3.8–13 cm)

Size
5.5–6.5 in (14–17 cm)

NORTHERN FLICKER
Colaptes auratus

male

female

    This attractive woodpecker,
which can be seen throughout the
region, is common in open woodlands and right into
suburban areas. Unlike most woodpeckers, it spends
some of its time feeding on the ground.

    A clear flicker track shows a distinctive arrangement
of two strong toes pointing forward and two pointing to
the rear, with the outer toes slightly longer than the inner
ones. The flicker's toes—along with short, strong legs that
give the bird a short stride—are well suited for grasping
tree trunks and limbs as this agile bird works its way
along in search of insects.

Similar Species: Most other birds have very different
tracks. Other woodpeckers would leave similar tracks,
but very few come down to the ground as much as the
obliging Northern Flicker does.

Dark-eyed Junco

Print
Length: to 1.5 in (3.8 cm)
Straddle
1–1.5 in (2.5–3.8 cm)
Stride
Hopping: 1.5–5 in (3.8–13 cm)
Size
5.5–6.5 in (14–17 cm)

DARK-EYED JUNCO
Junco hyemalis

This common small bird typifies the many small hopping birds found in the region. Each foot has three forward-pointing toes and one longer toe at the rear. The best prints are left in light snow or moist, loose dirt, although sometimes the toe detail is lost. The footprints may show some dragging between the hops.

A good place to study this type of print is near a bird-feeder. Watch the birds scurry around as they pick up fallen seeds, then have a look at the prints left behind. For example, the Dark-eyed Junco is attracted to seeds that chickadees (*Poecile* spp.) scatter as they forage for sunflower seeds in the birdfeeder. Also look for tracks under coniferous trees, where juncos feed on fallen seeds in winter.

Similar Species: The Northern Cardinal (*Cardinalis cardinalis*) makes very similar tracks, as do many other small birds. Toe size may help with identification—larger birds make larger prints—as can the season. Mouse (pp. 84–87) tracks might appear similar, so follow the trail to see if it disappears down a hole or into thin air.

Frogs

fore

hind

Straddle
to 3 in (7.5 cm)

hopping

FROGS

Bullfrog

There is a diversity of frogs in the region—too many to list—all of which may leave tracks. The Spring Peeper's (*Pseudacris crucifer*) 1.5-inch (3.8-cm) length and its preference for thick undergrowth and shrubs near the water make its tracks a rare sight. The larger Cope's Gray Treefrog (*Hyla chrysoscelis*) spends most of its time in trees, coming down to breed and sing at night. The Green Frog (*Rana clamitans*) and the Pickerel Frog (*R. palustris*) both favor slow-moving, shallow water and swampy areas and are fairly widespread. The beautiful and widespread Southern Leopard Frog (*R. sphenocephala*), up to 5 inches (13 cm) long, makes larger tracks. Unusually large tracks are surely from the robust Bullfrog (*R. catesbeiana*), North America's largest frog, which can reach 8 inches (20 cm) in length.

The best place to look for frog tracks is along the muddy fringes of waterbodies. A frog's hopping action results in its two small forefeet registering in front of its long-toed hind prints. Frog tracks vary greatly in size, depending on species and age. Toads (p. 116) usually walk, but they may also hop.

Toads

hind   *fore*

Straddle
to 2.5 in (6.5 cm)

walking

116

TOADS

American
Toad

There are fewer toad species than frog species in Kentucky and Tennessee. The toad most likely to be encountered, and the most widespread, is the American Toad (*Bufo americanus*), which lives in many different moist habitats. Fowler's Toad (*B. fowleri*)—which some people consider to be a subspecies of Woodhouse's Toad (*B. woodhousii*)—is found scattered throughout the region in temporary pools and ditches. The Eastern Spadefoot (*Scaphiopus holbrookii*) lives in arid and semi-arid habitats throughout much of the region. Toads in this region can grow up to 5 inches (13 cm) long.

Undoubtedly, the best place to look for toad tracks is, as with frog tracks, along the muddy fringes of water-bodies, but they can occasionally be found in drier areas—for example, as unclear trails in dusty patches of soil. In general, toads walk and frogs (p. 114) hop, but toads are pretty capable hoppers, too, especially when being hassled by overly enthusiastic naturalists. Toads leave rather abstract prints as they walk. The heels of the hind feet do not register. On less firm surfaces, the toes often leave draglines.

Salamanders & Newts

fore

hind

Straddle
to 4 in (10 cm)

walking

SALAMANDERS & NEWTS

Eastern Tiger Salamander

There are a wealth of salamanders and newts in the moist and wet areas of Kentucky and Tennessee. Among the more abundant and widespread of these long, slender, lizard-like amphibians is the Eastern Newt (*Notophthalmus viridescens*), which can grow to 5.5 inches (14 cm) in length. After a fresh rain, Eastern Newts emerging from ponds may leave small trails in the mud.

The Spotted Salamander (*Ambystoma maculatum*), which can grow to 10 inches (25 cm) in length, lives throughout the region in mixed forests and some coniferous forests. The king of the salamander world is undoubtedly the magnificent Eastern Tiger Salamander (*A. tigrinum*), which can grow up to 13 inches (33 cm) long and comes in such a diversity of colors and patterns that it defies description. This heavy salamander, found throughout Tennessee and in western Kentucky, leaves the best tracks, with a straddle of up to 4 inches (10 cm).

In general, a salamander fore print shows four toes, and the larger hind print shows five. However, print detail is often blurred by the animal's dragging belly or by the swinging of its thick tail across the tracks.

Lizards & Skinks

fore

hind

Straddle
to 3 in (7.5 cm)

walking

LIZARDS & SKINKS

Five-lined Skink

Lizards can be locally abundant in Kentucky and Tennessee, and in some areas they can be more common than amphibians. Lizard tracks are similar to salamander tracks, but they show longer, more slender toes. If you find lizard tracks, a likely candidate is the Five-lined Skink (*Eumeces fasciatus*), which can grow to 8 inches (20 cm) long. It favors moist woodlands and is widespread throughout the region. Another skink that you might encounter is the Broadhead Skink (*E. laticeps*).

The Racerunner (*Cnemidophorus sexlineatus*) is also a likely suspect, although it is less common in Kentucky. This lizard, which can reach 11 inches (28 cm) in length, favors dry grasslands and well-drained woodlands. The Fence Lizard (*Sceloporus undulatus*), which can grow to 8 inches (20 cm) in length, is common in open woodlands and grasslands.

These reptiles all move very quickly when the need arises, their feet barely touching the ground as they dart for cover. Consequently, their tracks can be hard to make out clearly.

Turtles

fore

hind

fore

hind

Straddle
4–10 in (10–25 cm)

*Snapping Turtle
walking*

*typical turtle
walking*

TURTLES

Painted Turtle

Turtles, those ancient inhabitants of the water world, will happily slip into the murky depths to avoid detection, but they do come out from time to time to feed or to bask in the sunshine. Look for their distinctive tracks alongside ponds, rivers and moist areas. Some turtles, such as the huge Common Snapping Turtle (*Chelydra serpentina*), rarely come onto land. One of the most widespread and prettiest turtles, often seen basking, is the Painted Turtle (*Chrysemys picta*); it can reach almost 10 inches (25 cm) in length. Slightly smaller is the well-named Common Musk Turtle (*Sternotherus odoratus*), which lives in shallow water and produces a foul odor. The Slider (*Trachemys scripta*), which reaches 12 inches (30 cm) in length, prefers slow-moving rivers. The Eastern Box Turtle (*Terrapene carolina*) is common in moist, forested areas and also in grasslands.

With its large shell and short legs, a turtle leaves a track pattern that has a straddle about half its body length. Although longer-legged turtles can raise their shells off the ground, short-legged species may let them drag, which is shown in their tracks. The tail may leave a straight dragline in the mud. On firmer surfaces, look for distinct claw marks.

Snakes

SNAKES

Common Garter Snake

Many snake species inhabit the region, and they thrive in both arid and moist conditions. Because all snakes are long and slender, their tracks appear so similar that identification among the species is next to impossible. In fact, because a snake lacks feet and leaves a track that is just a gentle meander, it is very challenging to establish in which direction a snake was moving.

The most widespread, frequently encountered snake is the harmless Common Garter Snake (*Thamnophis sirtalis*). Found throughout the region, often close to wet or moist areas, it can reach 4.3 feet (1.3 m) in length. Also widespread is the Redbelly Snake (*Storeria occipito-maculata*), which prefers hilly woodlands and can grow to 16 inches (40 cm) long. The Rough Green Snake (*Opheodrys aestivus*), which can grow to 3.8 feet (1.2 m) in length, spends most of its time in bushes and low-hanging branches near water. The rattlesnake most likely to be encountered is the Timber Rattlesnake (*Crotalus horridus*). Growing to 6.3 feet (1.9 m) in length, it frequents a variety of habitats from marshlands to dry woodlands. Also common in a variety of habitats is the large and colorful Milk Snake (*Lampropeltis triangulum*).

TRACK PATTERNS & PRINTS

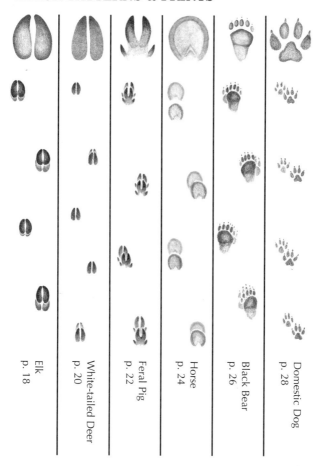

Elk
p. 18

White-tailed Deer
p. 20

Feral Pig
p. 22

Horse
p. 24

Black Bear
p. 26

Domestic Dog
p. 28

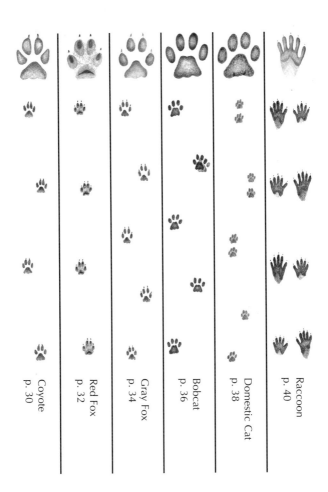

Coyote
p. 30

Red Fox
p. 32

Gray Fox
p. 34

Bobcat
p. 36

Domestic Cat
p. 38

Raccoon
p. 40

TRACK PATTERNS & PRINTS

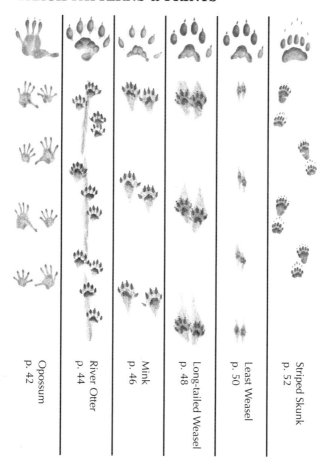

Opossum
p. 42

River Otter
p. 44

Mink
p. 46

Long-tailed Weasel
p. 48

Least Weasel
p. 50

Striped Skunk
p. 52

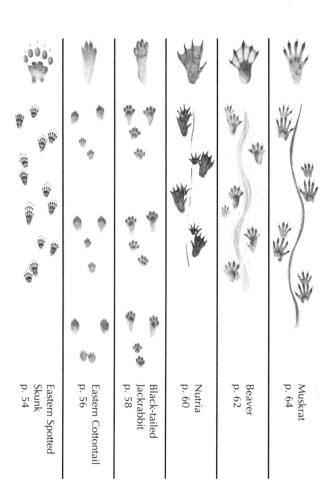

Eastern Spotted
Skunk
p. 54

Eastern Cottontail
p. 56

Black-tailed
Jackrabbit
p. 58

Nutria
p. 60

Beaver
p. 62

Muskrat
p. 64

129

TRACK PATTERNS & PRINTS

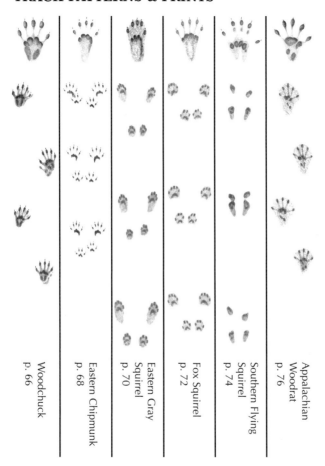

Woodchuck
p. 66

Eastern Chipmunk
p. 68

Eastern Gray
Squirrel
p. 70

Fox Squirrel
p. 72

Southern Flying
Squirrel
p. 74

Appalachian
Woodrat
p. 76

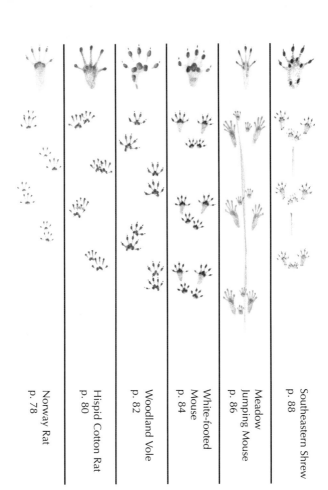

TRACK PATTERNS & PRINTS

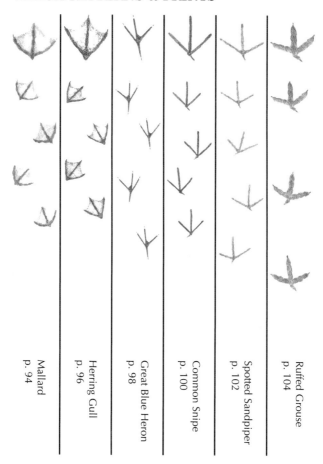

Mallard
p. 94

Herring Gull
p. 96

Great Blue Heron
p. 98

Common Snipe
p. 100

Spotted Sandpiper
p. 102

Ruffed Grouse
p. 104

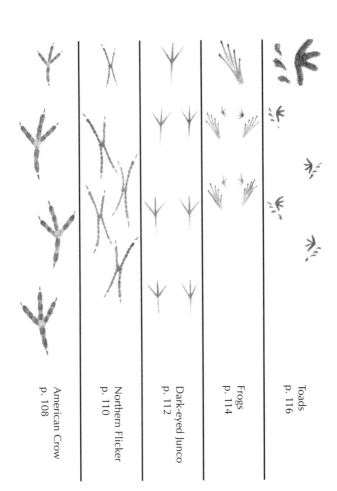

American Crow
p. 108

Northern Flicker
p. 110

Dark-eyed Junco
p. 112

Frogs
p. 114

Toads
p. 116

TRACK PATTERNS & PRINTS

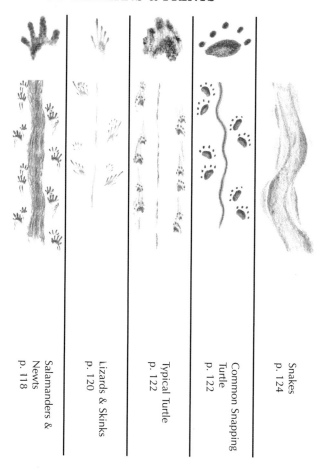

Salamanders &
Newts
p. 118

Lizards & Skinks
p. 120

Typical Turtle
p. 122

Common Snapping
Turtle
p. 122

Snakes
p. 124

HOOFED PRINTS

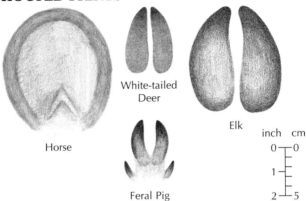

Horse

White-tailed
Deer

Feral Pig

Elk

inch   cm

HIND PRINTS

White-footed
Mouse

Woodland
Vole

Hispid
Cotton Rat

Southern
Flying Squirrel

Eastern
Chipmunk

Meadow
Jumping Mouse

inch   cm

HIND PRINTS

Norway
Rat

Appalachian
Woodrat

Southeastern
Shrew

Eastern
Cottontail

Woodchuck

Muskrat

Fox
Squirrel

Eastern Gray
Squirrel

inch   cm
0 — 0
1
2 — 5

Opossum

Raccoon

Black-tailed
Jackrabbit

HIND PRINTS

inch   cm
0 — 0
2
4 — 10

Nutria

Beaver

Black Bear

136

FORE PRINTS

Least
Weasel

Long-tailed
Weasel

Mink

River
Otter

Eastern
Spotted Skunk

Domestic
Cat

Gray
Fox

Striped
Skunk

Bobcat

Red
Fox

Coyote

Domestic Dog

inch   cm

0 — 0

1

2 — 5

137

BIBLIOGRAPHY

Behler, J.L., and F.W. King. 1979. *Field Guide to North American Reptiles and Amphibians.* National Audubon Society. New York: Alfred A. Knopf.

Brown, R., J. Ferguson, M. Lawrence and D. Lees. 1987. *Tracks and Signs of the Birds of Britain and Europe: An Identification Guide.* London: Christopher Helm.

Burt, W.H. 1976. *A Field Guide to the Mammals.* Boston: Houghton Mifflin Company.

Farrand, J., Jr. 1995. *Familiar Animal Tracks of North America.* National Audubon Society Pocket Guide. New York: Alfred A. Knopf.

Forrest, L.R. 1988. *Field Guide to Tracking Animals in Snow.* Harrisburg: Stackpole Books.

Halfpenny, J. 1986. *A Field Guide to Mammal Tracking in North America.* Boulder: Johnson Publishing Company.

Headstrom, R. 1971. *Identifying Animal Tracks.* Toronto: General Publishing Company.

Murie, O.J. 1974. *A Field Guide to Animal Tracks.* The Peterson Field Guide Series. Boston: Houghton Mifflin Company.

Rezendes, P. 1992. *Tracking and the Art of Seeing: How to Read Animal Tracks and Signs.* Vermont: Camden House Publishing.

Stall, C. 1989. *Animal Tracks of the Rocky Mountains.* Seattle: The Mountaineers.

Stokes, D., and L. Stokes. 1986. *A Guide to Animal Tracking and Behaviour.* Toronto: Little, Brown and Company.

Wassink, J.L. 1993. *Mammals of the Central Rockies.* Missoula: Mountain Press Publishing Company.

Whitaker, J.O., Jr. 1996. *National Audubon Society Field Guide to North American Mammals.* New York: Alfred A. Knopf.

INDEX

Page numbers in **boldface** type refer to the primary (illustrated) treatments of animal species and their tracks.

ABOUT THE AUTHORS

Tamara Eder, equipped from the age of six with a canoe, a dip net and a note pad, grew up with a fascination for nature and the diversity of life. She has a degree in environmental conservation sciences and has photographed and written about the biodiversity in Bermuda, the Galapagos Islands, the Amazon Basin, China and Southeast Asia.

Ian Sheldon, an accomplished artist, naturalist and educator, has lived in South Africa, Singapore, Britain and Canada. Caught collecting caterpillars at the age of three, he has been exposed to the beauty and diversity of nature ever since. He was educated at Cambridge University and the University of Alberta. When he is not in the tropics working on conservation projects or immersing himself in our beautiful wilderness, he is sharing his love for nature. Ian enjoys communicating this passion through the visual arts and the written word.